A CAT'S LIFE!

Photographs by Jane Burton

vmb
PUBLISHERS

Editorial Project: Valeria Manferto De Fabianis - Graphic Design: Clara Zanotti

Preface: Kim Taylor - Editorial Assistant: Giorgio Ferrero

A CAT'S LIFE!

Photographs by Jane Burton

CONTENTS

Preface by
KIM TAYLOR

A cat has a lithe and supple body. It can bound at speed across open ground or creep stealthily unseen in long grass. With its hooked claws it can climb swiftly up the bare trunk of a tree and leap unerringly from branch to branch. In the unlikely event of a fall, a cat can twist its body around in a trice so that it always lands on its feet. This athletic ability has been inherited from the cat's wild ancestors - and domestic cats are not far removed from wild cats.

The world is different...

...through the eyes of a cat

Selective breeding has produced a variety of coat colors but the basic shape and size of the wild ancestor have been retained.

Compare this with dogs, where selective breeding has produced a huge range in size and body form. The shapes and sizes of dogs have been bred for specific purposes because dogs are the servants of humankind. Cats on the other hand, are kept as companions, not servants, and though they enjoy human company, they are equally happy on their own.

A cat is a born hunter. The instinct to pounce on any small moving object is evident in kittens from almost as soon as they are able to walk. Kittens learn to hunt by playing with toys. Their games involve many variations of "stalk and pounce." But when the prey (toy) has been caught, a kitten will then play with it, tossing it and batting it with its forepaws. Of course, kittens also play with each other in the most delightful way. In the rough and tumble of play fighting, kittens sharpen their athletic skills. Even mature adult cats will sometimes indulge in play fighting. This may be to gain practice for life in the real world where fights may be serious. The domestic cat's wild ancestors were solitary animals, each defending its own territory, and the ability to fight off intruders was of primary importance.

Photographing cats in the home or garden can produce some beautiful images of the animals relaxed and at ease, but it is difficult by this means to show the grace and elegance of a cat's movement. To obtain detailed photos of cats in action it is necessary to use artificial flash lighting so as to avoid unacceptable blurring of the image. The flashlight on most modern cameras will only produce flat front lighting with harsh shadows. A photographic studio produces soft, all-round lighting, which compliments the curving body shape of the subject. But bringing a cat into a photographic studio presents an immediate problem. In unfamiliar surroundings, the cat may well crouch in a corner and refuse to cooperate. Unlike a dog, a cat cannot be trained to do something it does not want to do, and any amount of coaxing may well be to no avail. The answer is time. Given a day - or perhaps several - living in the studio, the cat begins to feel at home and will start to purr and play so that photos of a happy and relaxed cat can be obtained.

IN THE FAMILY

A game of relations…

…and generations

Affectionate lunges or rebellious instincts. Blind trust or need for independence. It's role playing. Or maybe it's just a typical day in the life of a family of cats…

Family Portrait

"Bella" and her kittens

PRIVATE TRANSPORTATION

"Pansy," Ticked tabby-tortoiseshell, with kitten

"Ozzie," Red Burmese, with kitten

Cuddle Time

THE REBEL

Red tabby British Shorthair

"Pansy," Ticked tabby-tortoiseshell, with kitten

Fur against Fur

"Phoebe," Blue-cream British Shorthair, and "Skylark"

MOTHERLY REPRIMANDS

"Aster," Silver spotted tabby, and "Asphodel"

"Kami," Korat, and her kitten

Sooner or later...

. . . your sins will find you out

"Kami," Korat, and her kitten

"Rasha," Brown spotted Bengal, and "Spike"

Pleasant Tickles

"Zap" and "Zelda," Silver tabby British Shorthair

The cat is a domestic tiger

Silver tabby cat and "Aster," silver tabby with kitten

A cat does not ask to receive;
it simply takes

"Aster," Silver spotted tabby, with kitten

LIFE LESSON

I understand that you love me!

Silver tabby

"Glenda." Red tabby, with kitten

Goodnight Kiss

"Lucky," domestic Shorthair Ginger, and "Hamish"

On a Trip with Mom

"Cookie" with its kitten and tortoiseshell female cat with kitten

“Cats live in worlds
we can only dream of”
(G. Priscilli)

NEED FOR INDEPENDENCE

DAVID AND GOLIATH

Tabby Burmese-cross and Silver tabby kitten

High Voltage

An Ambush...in the Family

"Sylvia," Silver tortoiseshell British Shorthair, with kitten

"There is not much less vexation in the government of a private family than in the managing of an entire state"

(Michel de Montaigne)

Come out to Play

"Pearl," Silver tortoiseshell-and-white, with kitten

"Bella," Blue-Agouti Burmese, with kitten

MOTHER'S ATTENTIONS

ADULT LIFE

The thousand faces of a cat...

...many-sided and mysterious masters of the house

Mysterious, elegant, intelligent, affectionate, independent, intrepid, playful...simply cats.

"Rosie" and "Salem," Black cat

ABSOLUTE BLACK

“Chiba,” Korat

NOBLESSE OBLIGE

The Odd Couple

"Felix" and "Felix," Black and white cat

Seal-point Siamese

Aerial Perspectives

The Enigma of the Sphinx

"Cosmos," Silver tabby chinchilla Persian

Silver tabby

Who Turned the World Upside Down?

"Levi." Blue-and-white Burmese-cross

BIG GAME HUNT

"Pansy," Tabby-tortoiseshell, and "Monty," Blue Burmese

THE GODFATHER

"Pansy," Tabby-tortoiseshell, and "Monty," Blue Burmese

TERRITORIAL DISPUTES

Knockout

"Dainty," Tabby, and "Mowgli"

Silver tabby

THE DUELISTS

Position of Strength

"The cat, even though domesticated, will always remain an independent, wild, little panther"
(Konrad Lorenz)

"Spike," Bengal

"I've met many thinkers and many cats,
but the wisdom
of cats is infinitely superior"

(Hippolyte Taine)

"Cobweb," Blue bicolor Persian

MAKE UP

"Cobweb," Blue bicolor Persian

Devon Rex

"The eyes of cats are an ocean of mysteries"
(Marcel Reney)

Simply a cat

Silver tabby

“Cynthia,” Silver tabby

“The real objection to
the great majority of cats
is their insufferable air
of superiority”

(Pelham Grenville Wodehouse)

CLASSIC...

... CASUAL

"Holly," Silver Egyptian Mau, and Sepia-spotted Bengal

Chocolate Burmese-cross and "Horace," silver tabby chinchilla

POISE COURSE

A Long and Boring Day

"Cosmos," Silver tabby chinchilla Persian

"Del," Blue Tonkinese

"There's no need for sculptures
in a house where a cat lives"

(Anonymous)

"...please don't sink your hidden claws"
(John Keats)

"Del," Blue Tonkinese

"Butterfly" and "Joan," two Silver tortoiseshells

Natural Instinct

Lunge and Defense

"Peregrine" and "Butterfly"

DIFFERENT ANGLES

"Lucky," Ginger tabby

Beauty Farm

READY, STEADY...

... GO!

Siamese and Brown classic tabby

Yoga Position

Ginger and "Faber," Cornish Rex

"Dainty," Burmese-cross tabby

“A cat's idea of comfort is absolutely incomprehensible to humans”

(Colette)

Brown spotted tabby

STRETCHING

Seduction in a Single Glance

"Few animals can show their
temper trough distinct expressions
as cats can do"

(Konrad Lorenz)

"Maudi Blacknose," tortoiseshell and white

READY TO ATTACK

"Maudi Blacknose," tortoiseshell and white cat

Ginger and "Horatio," British Shorthair

WATCH OUT FOR THOSE TWO!

"The cat signs
its intentions
with its tail"

(R. Gómez de la Serna)

LITTLE FELINES

Tales of kittens...

...facing their first challenges

"There is no need to teach cats how to have fun, since they are blessed with an inherent talent in this art" (James Mason)

FIRST STEPS

Brown tabby, Silver striped tabby, Brown tabby British Shorthair

Rude Awakening

Cream cat and Red tabby British Shorthair

NURSERY SCHOOL

Two week old kittens

"Taz," Blue British Shorthair

Perpetual Motion

Break Dance

Upside Down

"Thomasina," Lilac-tortoiseshell Persian-cross

RELAX

Silver tabby and Somali

BULLY...

... SHY

Tiptoeing Around

New Discoveries

"Lukie"

"Butterfly," Silver tabby

Ambush Rehearsals

Morning Gymnastics

Ginger kitten

BEAUTY SECRETS

Tortoiseshell kitten and grey-and-white kitten

The other side of the coin

"Come, superb cat, to my amorous heart; hold back the talons of your paws, let me gaze into your beautiful eyes of metal and agate"

(Charles Baudelaire)

Seal-point Siamese and Lilac-tortoiseshell Persian-cross

Let's Sort Out this Tangle...

Antistress

"Punky," Blue-cream

Blue tabby and Silver tabby

The Pleasures of Life

"Squirrel-tail" and Black-and-white

Waiting for the First Move

CROSSED CHALLENGES

Tabby kittens

Test of Strength

CHECKMATE

Cream and tortoiseshell

VICTOR...

Two kittens fighting

AND THE DEFEATED

Brown tabby and "Eyebright," Snowshoe

FIRST ROUND

Red tabby British Shorthair

War and Peace

Red tabby British Shorthair

CHARGE!

Actors Studio

Red tabby British Shorthair

Dancing...

...Dancing

FRIENDS AND ENEMIES

Close encounters...

...so different, yet so close

The chance encounter or a forced cohabitation with another animal is never a trivial matter for the cat. It forces him to face up to his own feline nature, with his deeper instincts and his insatiable curiosity. It's an explosive combination, an unpredictable mix which can lead to tender friendships as well as relationships of reciprocal mistrust. Especially if lifelong enemies are on the other side: dogs.

Do We Know Each Other?

Kitten with house sparrow

An Invitation to the Opera

Unbelievable Things...

Silver tabby with hedgehog and turtle

Silver tortoiseshell-and-white Maine Coon with guinea pig

Unexpected Outpourings of Affection

Tabby kitten with squirrel

Blue tabby with Cavalier King Charles Spaniel puppy

What Goes Around Comes Around...

A Matter of Respect

Kitten with Pembroke Welsh Corgi

Divergent opinions

Kitten tabby with Golden Retriever puppy

Blue-cream with Golden Cocker Spaniel

EQUAL EXCHANGE

ASPIRING TWINS

Blue tabby kitten with Golden Retriever puppy and black-and-white kitten with Border Collie puppy

OPPOSITES ATTRACT

British Shorthair Brown tabby with rabbit

Mind Reading

Fur Brothers

"Sabrina," Ginger, with rabbit

British Shorthair red spotted with two Border Collie puppies

A Good Friend is Worth his Weight in Gold

Ginger kitten with Basset Hound

Seal-point Siamese with Brown Roan Spinone

Survival of the Fittest

FACE TO FACE

Ginger kitten and puppy

Ginger kitten with Doberman Pinscher

Yin and Yang

The Pillars of Hercules

Tabby kitten with Bulldog

AN IMPETUOUS FRIEND

White cat with two Dalmatians

POLKA DOT PILLOWS

vmb
PUBLISHERS
VMB Publishers® is a registered trademark
property of Edizioni White Star s.r.l.

Via Candido Sassone, 24
13100 Vercelli, Italy
www.whitestar.it

Translation: Maddalena Neale

ISBN: 978-88-540-1508-1
1 2 3 4 5 6   14 13 12 11 10

Printed in China

ALL PHOTOGRAPHS ARE BY JANE BURTON/WARREN PHOTOGRAPHIC EXCEPT THE FOLLOWING:
PAGE 102, 127 AND 129 MARK TAYLOR/WARREN PHOTOGRAPHIC
PAGE 163 DK LIMITED/CORBIS